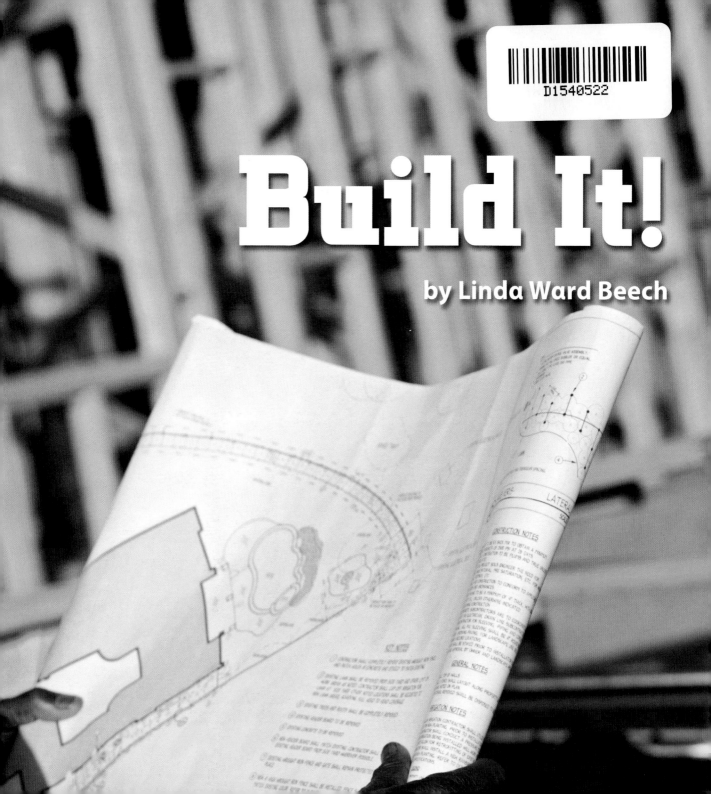

Build It!

by Linda Ward Beech

Scrape, scrape, scrape.
Bulldozers rumble.
They push rocks.
See them tumble!

push

3

Swoosh-a-swoosh.
Cranes go high.
They pull the beams
into the sky.

pull

push

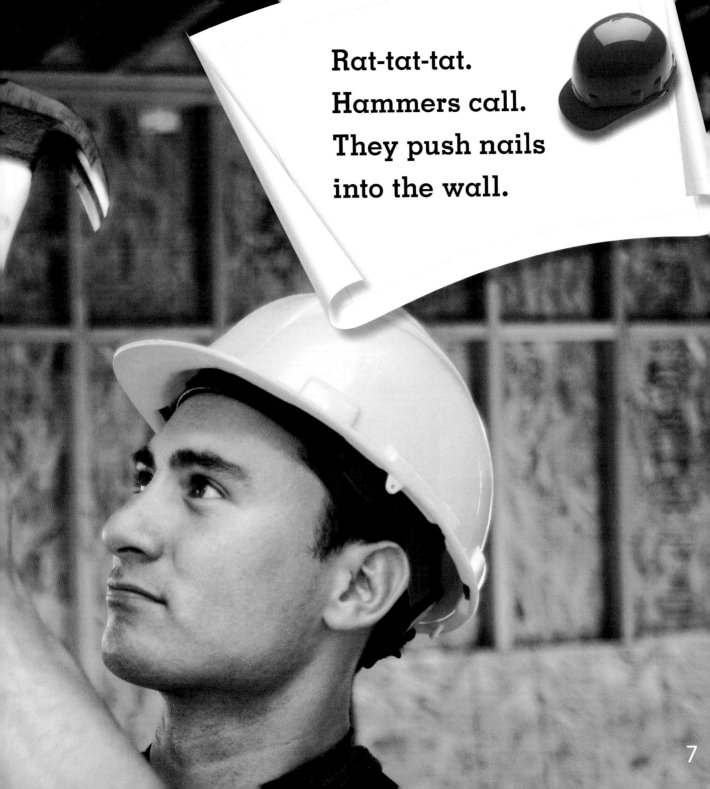

Rat-tat-tat.
Hammers call.
They push nails
into the wall.

Screech-a-screech.
Backhoes go deep.
They pull the dirt
into a heap.

8

pull

push

Drum, drum, drum.
Jackhammers sound.
They push down
into the ground.

Push and pull
to lift and dig.
Forces help
build something big.